BRAIN AEROBICS

MATH
PUZZLES

DERRICK NIEDERMAN

D0913286

PUZZLE
WRIGHT
PRESS
New York

PUZZLE
WRIGHT
PRESS
New York

An Imprint of Sterling Publishing
387 Park Avenue South
New York, NY 10016

© 2001 by Derrick Niederman

First Puzzlewright Press edition published in 2013.
This book was previously published under the title
Hard-to-Solve Math Puzzles.

ISBN 978-1-4549-0964-4

Distributed in Canada by Sterling Publishing
℅ Canadian Manda Group, 165 Dufferin Street
Toronto, Ontario, Canada M6K 3H6
Distributed in the United Kingdom by GMC Distribution Services
Castle Place, 166 High Street, Lewes, East Sussex, England BN7 1XU
Distributed in Australia by Capricorn Link (Australia) Pty. Ltd.
P.O. Box 704, Windsor, NSW 2756, Australia

For information about custom editions, special sales, and premium and
corporate purchases, please contact Sterling Special Sales
at 800-805-5489 or specialsales@sterlingpublishing.com.

Manufactured in the United States of America

2   4   6   8   10   9   7   5   3   1

www.puzzlewright.com

Contents

Introduction

Good news! You don't need to know any higher math to do the puzzles in this book. There are no prerequisites of differential calculus, functional analysis, linear algebra, or anything like that. But while the puzzles don't require advanced mathematics to state, they do require insight to solve: How can you divide a regular pentagon into five identical pentagonal shapes? What is the only decade in American history to contain four prime-numbered years? What three right triangles with integer sides have areas numerically equal to twice their perimeters?

If you don't see the answers just yet, don't give up. Remember, you have plenty of company. But if you come up with the right insight and solve the puzzles, you'll have a satisfaction you won't forget. This is one book that you'll be proud to carry around.

This book wouldn't have been possible without the work of some giants in the field. The author extends his thanks to everyone who showed how much fun puzzles are, from 19th-century greats Sam Loyd and Henry Dudeney all the way to Martin Gardner. The work of Joseph Madachy, Leo Moser, Harry Nelson, Arlet Ottens, Richard Stanley, and other top-notch puzzle makers and solvers has also had a great influence. Rob Blaustein suggested some interesting puzzle concepts. Fraser Simpson provided timely and thorough proofreading. Last but not least, my editor Peter Gordon put everything together into a tidy little book. Thanks to all.

—Derrick Niederman

Puzzles

1. WHEN IN ROME

Equations of the following sort are called either *cryptarithms* or *alphametics*. Whatever you want to call them, the idea is to substitute a number for each letter so that the indicated arithmetic is correct. No two distinct letters can be given the same number, and once a number has been substituted for a letter, it must substitute for each appearance of that letter. Also, no letter on the left side of any number can represent 0.

The following alphametic isn't particularly difficult, but it is special because the equation is true in Roman numerals; that is, $44 \times 10 = 440$. Your challenge is to make the equation true with regular numbers as well. There are two solutions.

$$\begin{array}{r} \text{X L I V} \\ \times \qquad\quad \text{X} \\ \hline \text{C D X L} \end{array}$$

2. FIND THE SHORTCUT

We know that $5^3 = 125$ and $6^3 = 216$. With that in mind, suppose you were told that the number 148,877 is the cube of some other whole number. What would that other number be?

(You don't need a calculator to do this problem: Once you find the right track, it's simpler than it first appears. It doesn't look simple, though, does it?)

3. PAGE BOY

As Jack finished a section of the novel he was reading, his wife posed a curious question: "Dear, suppose you took the page numbers of the section you just read and added them together. What would you get?"

Jack summed them rather hastily and said, "Well, I may have to double-check my arithmetic. It's either 412 or 512, I'm not sure."

Which was it?

4. CIRCULAR LOGIC

The diagram below features four houses, labeled C, X, Y, and Z. One of the houses—C—is located at the center of a circle, and the other three—X, Y, and Z—are somewhere on the circumference. There are lines connecting various houses, as shown.

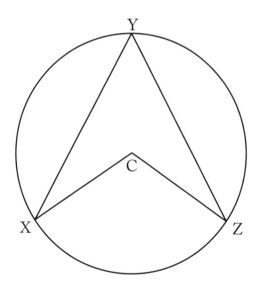

A member of family Y decides to take a walk starting from his house going clockwise around the circular path. Meanwhile, another member of family Y walks along the straight path, visiting houses X, C, and Z in order—but not stopping—before returning home.

Assuming that the two members of family Y walk at the exact same rate, who arrives home first?

5. ACROSS TO BEAR

A crossword puzzle was found torn, with only the top half intact. By looking at this piece alone, and knowing that crosswords are rotationally symmetric about the center square, it is possible to determine the number of the puzzle's final across entry. Can you figure out this missing number?

6. NO CALCULATORS, PLEASE

Prove the following inequality without multiplying the whole mess out.

$$(½) \times (¾) \times (⅚) \times (⅞) \times \ldots \times (⁹⁷/₉₈) \times (⁹⁹/₁₀₀) < ¹/₁₀$$

7. BORN UNDER A BAD SIGN

Sue Perstition was born on Friday the 13th. She recently celebrated one of the birthdays below on Friday the 13th as well. Which one was it?

    A) 10th
    B) 20th
    C) 30th
    D) 40th
    E) 50th
    F) 60th

8. CROWD O' THREES

Note that $2 = 3 - \frac{3}{3}$ and $6 = 3! + 3 - 3$, where 3! equals $3 \times 2 \times 1 = 6$.

Can you represent each of the numbers from 1 through 20 by using *precisely* three 3's? Only standard mathematical symbols are permitted. For good measure, come up with alternative representations of 2 and 6 than those presented above.

9. TENNIS, ANYONE?

Eight players entered the annual round-robin cash tournament at the Playfair Tennis Club. Over the course of the weekend, each player played one set against each of the other seven players. After the matches were over, players received the same number of dollars per set won as the total number of sets won. For example, if a player won five sets, he would receive five dollars for every set won, for $25 altogether. If a player won just one set, he would receive just one dollar, period.

The club collected an entry fee of $17.50 from each player, and the tournament organizers noted that this would always be enough to pay for the prizes. Why is this so?

10. ONE, TWO, THREE

Using all of the digits from 1 to 9 once each, create three 3-digit numbers that are in a ratio of 1:2:3. There are four solutions.

11. DON'T MAKE MY BROWN EYES BLUE

On a faraway island there was a kingdom in which the inhabitants had either blue eyes or brown eyes. It was a small island, with the property that with every passing day, everyone on the island saw everyone else. However, there were no mirrors or anything else that reflected, so no one knew the color of his or her own eyes. One day the king decreed the following: 1) At least one of you has blue eyes, and 2) if you wake up in the morning and realize you have blue eyes, you must leave the island at once, without letting anyone see you.

Well, it turns out that there were precisely ten people on the island with blue eyes. Given that everyone on the island was capable of perfect logical reasoning and that everybody on the island knew about the logical abilities of their fellow islanders, what do you suppose happened to them following the king's decree?

12. THE STAMP COLLECTION

Phil Atelist has a stamp collection consisting of three books. The first book contains ⅕ of the total number of stamps, the second book contains some number of sevenths of the total number of stamps (he can't remember how many), and the third book contains 303 stamps. How many stamps are in the entire collection?

13. TOO CLOSE TO CALL

Which is bigger? (No calculators allowed!)

$$\sqrt[10]{10} \quad \text{or} \quad \sqrt[3]{2}$$

14. THE LONG STRING

Are there ever 1,000,000 consecutive composite numbers?

15. FIRST-CLASS LETTERS

Define the "letter class" of a whole number as the number of letters in that number: For example, the letter class of 16 (SIXTEEN) is 7, the letter class of 25 (TWENTY-FIVE) is 10, and so on.

There is only one number between 1 and 5,000 that is the *only* representative of its letter class. Care to find it?

16. MOST VALUABLE PUZZLE

When the Sprocketball Writers Association of America votes on the annual Most Valuable Player award, three players are nominated and 10 writers designate a player as first, second, or third place, with each first-place vote worth three points, each second-place vote worth two points, and each third-place vote worth one point. Only the three nominated players can receive votes. (Somewhat different from the procedures followed by the Baseball Writers Association of America, but that's sprocketball for you.)

Under these conditions,

1) What is the smallest number of first-place votes that it would take to clinch the MVP?

2) What is the smallest number of first-place votes a candidate could receive and still win the MVP?

17. PERFORMANCE ANXIETY

Some years ago a leading brokerage firm touted its selections for the just-concluded year by pointing out that its group of "single-best" picks from its top analysts outperformed the S&P 500 by 32 percent. Given that the S&P 500 rose 28 percent for the year in question, that was no mean feat. Only problem was, the firm's selections didn't actually increase by 60 percent, as you might now be thinking. What do you suppose the actual performance was?

18. DOUBLE TROUBLE

Place the numbers one through nine in the boxes below so that *both* multiplications are true.

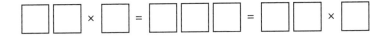

19. CEREAL SERIAL

A cereal company places prizes in its cereal boxes. There are four different prizes distributed evenly over all the boxes that the company produces. On average, how many boxes of cereal would you need to buy before you collected a complete set?

20. PAIRING OFF

It is easy to see that the ordered pair (3,2) satisfies the equation $x^2 - 2y^2 = 1$. But what is the next-smallest pair of positive integers that satisfies the same equation?

21. A SQUARE DEAL

A rich man bequeathed a square plot of land to his three children. The number of miles on each side of the plot was not recorded, but it was a whole number. To the first child, the man gave a square plot in the upper-right corner; again, its size was not recorded, but each side was a whole number. To his second child, he gave a plot that was square but for the upper-right corner that the first child owned; again, the sides of this plot were all whole numbers. The third child received the remaining plot, which turned out to be precisely equal in area to the second child's plot. The figure below shows what the plots looked like, although it is not necessarily to scale.

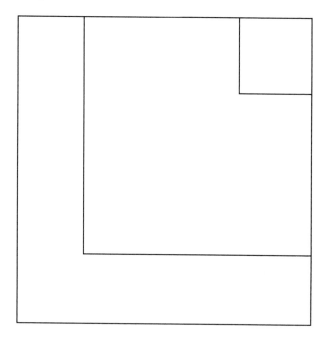

This information is not enough to determine the precise dimensions of each of the plots, but if you know that the first child received the smallest piece possible based on this information, you now have enough to figure out the dimensions of all three of the plots. How big were they?

22. SURVIVAL OF THE SPLITTEST

A population starts with a single amoeba. Suppose there is a ¾ probability that the amoeba will split to create two amoebas, and a ¼ probability that it will be unable to reproduce itself, in which case it will die out. Assuming that all future generations of amoebas have the same probabilities associated with them, what is the probability that the family tree of the single amoeba will go on forever?

23. SHUTTING THE EYE

The figure below may look somewhat familiar. Suppose the radius of the big circle is three times the radius of the small circle. Furthermore, suppose that the "eyelids" are made up of two separate arcs of an even larger circle. What is the length of the radius of that third circle relative to the "pupil" in the diagram?

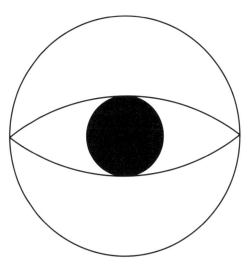

24. WHAT'S IN A NAME?

As you can see, each of the following three alphametics involves a famous name. Unfortunately, only one of these equations is possible. Can you determine which one has a solution? (For the basic rules of alphametics, see puzzle 1.)

$$
\begin{array}{r}
M A R Y \\
+ \ T Y L E R \\
\hline
M O O R E
\end{array}
$$

$$
\begin{array}{r}
J O Y C E \\
+ \ C A R O L \\
\hline
O A T E S
\end{array}
$$

$$
\begin{array}{r}
J A M E S \\
- \ E A R L \\
\hline
J O N E S
\end{array}
$$

25. THUMBS DOWN

It turns out to be impossible to "reverse" a number by multiplying it by 2. In other words, there is no number of the form abcd, for example, such that abcd × 2 = dcba. (The equation is not only impossible for four-digit numbers, it is impossible for *all* numbers.)

However, there is a three-digit number abc *in base 8* such that abc × 2 = cba. Can you find that number?

26. TICKET TO RIDE

Each of the railroad stations in a certain area sells tickets to every other station on the line. This practice was continued when several new stations were added, and 52 additional sets of tickets had to be printed. How many stations were there originally, and how many new ones had to be added?

27. GOING OFF ON A TANGENT

In the diagram below, a circle is inscribed in an isosceles trapezoid whose parallel sides have lengths 8 and 18, as indicated. What is the diameter of the circle?

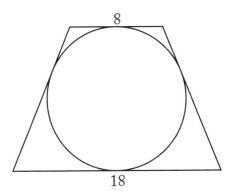

28. SURPRISE ENDING

For what integral values of n is the expression $1^n + 2^n + 3^n + 4^n$ divisible by 5?

29. THANKSGIVING FEAST

On the first Thanksgiving in Plymouth, Massachusetts, legend has it (a just-created legend, that is) that 90 percent of the pilgrims had turkey, 80 percent had corn, 70 percent had pumpkin pie, and 60 percent had mince pie. No one, however, had all four items. Given the limited menu, what percentage of pilgrims had at least one of the two desserts?

30. ALL ABOUT PYTHAGORAS

Recall that a Pythagorean triple is a set of three positive integers that can form the sides of a right triangle. The best known example is the famous 3–4–5 right triangle below:

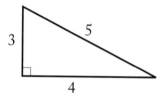

These dimensions create a right triangle because, like any self-respecting Pythagorean triple, they satisfy the Pythagorean theorem: $3^2 + 4^2 = 5^2$

(1) What positive integers cannot be part of a Pythagorean triple?

(2) What is the smallest number that can be used in all three positions—as the hypotenuse, as the longer leg, and as the shorter leg—in three different right triangles?

31. A DECADENT DECADE

Only once in American history has there been a decade in which four of the years were prime numbers. Can you find it?

32. COMPOSITE SKETCH

If the previous problem was too hard, you might want to tackle this one instead: Can you find the first decade (A.D., of course) in which all of the years were composite? To start you off on the right track, you should know that America was not around during this decade.

33. FIVE SQUARES TO TWO

Using just two straight cuts, divide the figure below into three pieces and reassemble those pieces to form a rectangle twice as long as it is wide.

34. EQUALITY, FRATERNITY

1) Consider the "equation" 6145 − 1 = 6143. Can you move two digits so as to create a valid equality?

2) The equation −127 = −127 is, of course, true. Suppose you were told to move two digits so as to leave a valid equation. You might try moving both 2's to the end, creating the equation −172 = −172, but you've probably guessed that you have to move two digits on the same side of the equation. Any ideas?

35. BLUE ON BLUE

A hat contains a number N of blue balls and red balls. If five balls are removed randomly from the hat, the probability is precisely ½ that all five balls are blue. What is the smallest value of N for which this is possible?

36. STICK FIGURES

Consider the following equation, which is obviously false.

$$1 \ 1 = 1 \ 1 \ 3 \ 3 \ 5 \ 5$$

A) Can you insert four line segments into this equation to make it correct?

B) Same question, but now you are given only three line segments to work with. To give you a break, the equation doesn't have to be exact, but it does have to be accurate to six decimal places!

37. HERALDING LOYD

Sam Loyd was perhaps the greatest puzzle maker in American history. One of his creations was an elegant puzzle that concerned crossing a waterway by boat. You should enjoy its beautiful simplicity.

Two ferryboats traveling at a constant speed start moving at the same instant from opposite sides of the Hudson River, one going from New York City to Jersey City and the other from Jersey City to New York City. They pass one another at a point 720 yards from the New York shore.

After arriving at their respective destinations, each boat spends precisely 10 minutes at the opposite shore to change passengers before switching directions. On the return trip, the two boats meet at a point 400 yards from the Jersey shore.

What is the width of the river?

38. LOTS OF CONFUSION

You have another chance to match wits with Sam Loyd. This Loyd original concerns the case of a real estate mogul who bought a piece of land for $243, divided it into equal lots, and sold the entire package for $18 per lot. (If the prices seem low, remember that Loyd lived in the 19th century.) The mogul's profit on the entire deal was equal to what six of the lots had originally cost him.

How many lots were in the piece of land?

39. OH, HENRY!

Speaking of famous puzzle makers and lots of land, Henry Dudeney (1847–1930) was one of the greatest puzzlists of all time, and the puzzle below is adapted from one of his many brilliant creations. Perhaps the biggest shock is that it can be solved without higher mathematics or even a calculator.

The diagram displays three square plots of land. Plot A is 388 square miles, plot B is 153 square miles, and plot C is 61 square miles. How big is the triangular plot of land between A, B, and C?

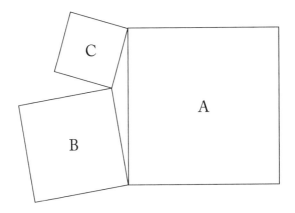

40. CATCH-22

You may recall that n factorial, written n!, is the product of all positive integers less than or equal to n. Well, there are five numbers n such that n! has precisely 22 zeroes at the end. Can you find these five numbers?

When you're finished with that one, can you also name the numbers whose factorials end with *23* zeroes?

41. WALKING THE BLANK

The number below is intended to be a 28-digit number, but ten of the digits have been left blank. These blanks are to be filled with the digits 0, 1, 2, 3, 4, 5, 6, 7, 8, and 9—which, for the record, can be done in 10! = 3,556,800 different ways. What is the probability that the resulting 28-digit number will be divisible by 396?

5_383_8_2_936_5_8_203_9_3_76

42. BREAKING THE HEX

If you start with a regular hexagon and join its vertices as shown, you will create a smaller regular hexagon in the middle. If the larger hexagon has an area of one square foot, what is the area of the smaller hexagon?

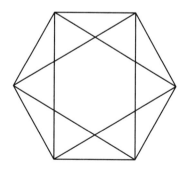

43. DOMINO THEORY

Suppose you had two dominoes that looked like this:

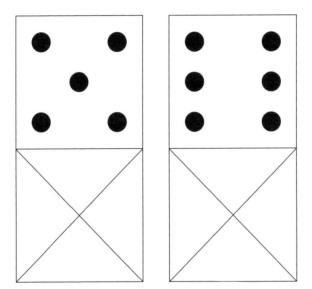

The X's at the bottoms of the dominoes mean that these areas are not visible, so all you know is that one of the dominoes has a five-spot and the other has a six-spot. What is the probability that you can form an end-to-end chain of all 28 dominoes—with the two depicted dominoes in first and last position—subject to the usual rule that the number on the right of any domino in the chain always equals the number on the left of the next domino?

44. THE WAYWARD THREE

There is an integer whose first digit is 3 having the property that if you take the 3 from the beginning of the number and place it at the end, you will have multiplied the original number by $\frac{3}{2}$. What is that number?

45. OH, REALLY?

Show that the following is true. (No calculators allowed!)

$$\sqrt[3]{2+\sqrt{5}} + \sqrt[3]{2-\sqrt{5}} = 1$$

It's simpler than it looks. Really!

46. DOWN TO THE WIRE

Suppose two evenly matched teams play in the World Series. They are so evenly matched that the probability of either team's winning any particular game is precisely 50 percent. Not only that, the teams don't get overconfident or discouraged, so the 50 percent probability doesn't change as the Series goes on.

Under these conditions, it is quite unlikely that either team will engineer a four-game sweep. In fact, it turns out that a sweep is precisely one-half as likely as the Series' ending in five games. What is the likelihood that the Series will go the full seven games?

47. EASY AS A, B, C

Can you find three distinct positive integers A, B, and C such that the sum of their reciprocals equals 1?

48. THE BEANPOT RALLY

Every winter in Boston, four area schools—Boston College (BC), Harvard, Northeastern, and Boston University (BU)—compete in a hockey tournament called the Beanpot. One year BU defeated Northeastern in the first round, while Harvard defeated BC. BU went on to defeat Harvard in the final, while BC defeated Northeastern for third place. The tournament draw looked like this:

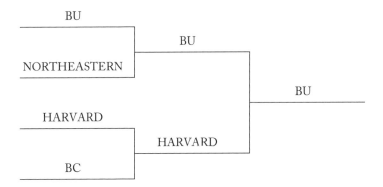

BU

BU

NORTHEASTERN

BU

HARVARD

BU

HARVARD

BC

If the above set of results—coupled with the third-place playoff—is considered one possible outcome for the tournament, how many possible outcomes are there in all?

49. 3, 4, 6, HIKE!

A hiker went on a little expedition. The first part of the trip was on level ground, and he walked at 4 miles per hour. The second part was uphill, and his speed slowed to 3 m.p.h. He then retraced his steps, going 6 m.p.h downhill and then the same 4 m.p.h. back to his starting point. Given that the total trip took five hours, how far did the hiker walk?

50. EVEN STEVEN

The one-digit odd numbers—1, 3, 5, 7, and 9—add up to 25, while the one-digit even numbers—0, 2, 4, 6, and 8—add up to 20. Your challenge is to arrange the numbers in such a way that the odd numbers and even numbers have the same value. You can use +, −, ×, ÷, and also combine digits to make multi-digit numbers.

51. SECOND MOST VALUABLE PUZZLE

This puzzle takes over where puzzle 16 left off, except now we suppose that there are only three sportswriters placing votes. We also suppose that the point awards for first-, second-, and third-place votes are not known precisely. All we know is that a first-place vote is worth x points, a second-place vote is worth y points, and a third-place vote is worth z points, with x > y > z.

When the points have been counted up, Player A has emerged with 20 points, Player B with 10 points, and Player C with 9 points. If Player A received a second-place vote from the *Daily News*, who received a second-place vote from the *Herald-Tribune*?

52. TOP SCORE

If a bunch of positive integers adds up to 20, what is the greatest possible *product* of these numbers?

53. A BRIDGE TOO FAR

In the game of duplicate bridge, the idea is to get a better score than the other pairs who, during the course of the session, play the same hands at different tables. A pair will get one point for every score they beat, and half a point for every score they tie.

One particular hand was played eight times. All pairs playing North-South scored either +450 or +420. (The derivation of these scores doesn't matter, although they are common results for a contract of, say, four hearts.) One of the pairs that scored +420 noted that this score was worth 2½ points at the end of the session. How many points would this pair have received had they scored +450 instead?

54. SEVEN-POINT LANDING

Plot seven points on a sheet of paper in such a way that if you choose any three of them, at least two will be precisely one inch apart. (Note: You don't have to bend the paper in any way. The points should all be in the same plane.)

55. 'ROUND GOES THE GOSSIP

One puzzle that is destined to become a classic (and is being mentioned here in order to improve its chances) is the "gossip" puzzle, which has several variations but goes essentially like this:

There are six busybodies in town who like to share information. Whenever one of them calls another, by the end of the conversation they both know everything that the other one knew beforehand. One day, each of the six picks up a juicy piece of gossip. What is the minimum number of phone calls required before all six of them know all six of these tidbits?

56. WHAT'S IN A NAME, PART TWO

We have moved from addition and subtraction (puzzle 24) to multiplication, but the question remains the same: What is the unique solution to the following cryptarithm? (Recall that you must substitute a number for each letter, with repetitions of the same letter to be given the same number.)

$$\begin{array}{r} \text{E D W A R D} \\ \times \qquad\quad \text{R} \\ \hline \text{M U R R O W} \end{array}$$

27

57. TWO SQUARES ARE BETTER THAN ONE

By cutting the checkerboard below into four pieces and adding a single square, it is possible to form two smaller square checkerboards. Good luck.

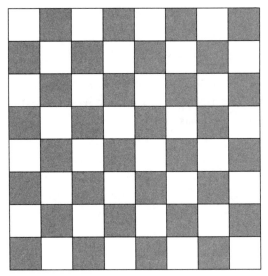

58. HITTER'S DUEL

Let's say that Ty Cobb's season batting average is the same as Shoeless Joe Jackson's at the beginning of a late-season doubleheader. (Assume both players have had hundreds of at bats.) Cobb went 7 for 8 on the day (.875), while Jackson went 9 for 12 (.750). But at the end of the day, Jackson's season average turned out to be higher than Cobb's. How is this possible?

59. VISIBLE AND DIVISIBLE

Can you replace the missing digits in the number 789,XYZ so that the resulting number is divisible by 7, 8, and 9? The only restriction is that you cannot use a 7, an 8, or a 9.

60. ONE PIECE FEWER

Is it possible to perform the dissection of problem 57 by cutting the 8 × 8 checkerboard into just *three* pieces instead of four?

61. SCALE DRAWING

The Celsius scale is derived from the Fahrenheit scale by making a linear adjustment: Specifically, whereas the freezing and boiling points of water are 32 degrees and 212 degrees Fahrenheit, the Celsius scale was created to make these important points 0 degrees and 100 degrees, respectively.

There is one temperature that reads the same on both the Fahrenheit and Celsius scales. What is it?

62. PLAYING THE TRIANGLE

Pictured below is an isosceles right triangle. Using three straight cuts, divide the triangle into four pieces that can be put together to create two smaller isosceles right triangles that are different sizes.

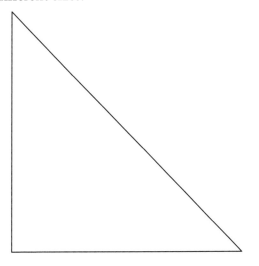

63. HOW BIG?

The figure below shows two circles of radius one, with the center of each lying on the circumference of the other. What is the area of the wedge-shaped region common to both circles?

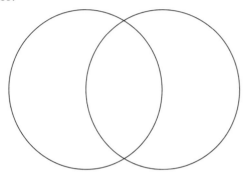

64. TUNNEL DIVISION

A train traveling at 90 miles per hour takes four seconds to completely enter a tunnel, and 40 additional seconds to completely pass through the tunnel. How long is the train? How long is the tunnel?

65. PRIME TIME

Arrange the digits 0, 1, 2, 3, 4, 5, 6, and 7 so that the sum of any two consecutive digits is a prime number. (Remember that 1 is not considered prime.) There is more than one possible answer.

66. TRIANGLE EQUALITIES

There are precisely three right triangles with integral sides having the property that their area is numerically equal to twice their perimeter. Can you find them?

67. DIVIDING THE PENTAGON

It is easy to divide an equilateral triangle into three equal (though not equilateral) triangles. It is even simpler to divide a square into four equal squares. These constructions are given below:

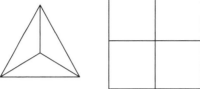

Now for the tough part. Can you divide the regular pentagon below into five equal pentagons? (The smaller pentagons will of course not be regular pentagons.)

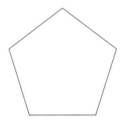

68. PLANTING THE SOD

It is well known that a number is divisible by 9 if and only if the sum of its digits is divisible by 9. For example, $3 + 9 + 6 = 18$ is divisible by 9, and $396 = 9 \times 44$. Let's introduce the SOD operator as one that sums the digits in a number, so that $SOD(396) = 18$.

With all this in mind, can you determine the value of $SOD(SOD(SOD(4444^{4444})))$?

One piece of advice: Please don't start calculating 4444^{4444}, as the world will likely end before you or your descendants ever finish the job.

69. HIGHER THAN YOU THINK

What is the smallest number N such that it is impossible to have $1.00 in change consisting of precisely N coins? You can use half-dollars, quarters, dimes, nickels, and pennies—and even a Sacajawea dollar for the case N = 1.

70. POCKET CHANGE

In the unlikely event that the previous problem was too easy, here's a genuine toughie for you along similar lines:

Suppose a friend of yours announces that he has a number of coins in his pocket that add up to precisely one dollar. When he tells you how many coins he has, you ask if any one of them is a half-dollar, and he answers no. You quickly realize that you can't tell for sure what coins he has, because there are six different combinations that produce precisely one dollar.

How many coins does your friend have in his pocket?

71. SQUARE NOT

Note first that the number 150 is expressible as the sum of distinct squares, as shown below.
$$150 = 100 + 49 + 1 = 10^2 + 7^2 + 1^2$$
We'll spare you the trouble, but you can take our word for it that every number greater than 150 is also expressible as the sum of distinct squares. But there are 37 numbers that cannot be expressed in this fashion. Care to find the largest one?

72. DUELING WEATHERMEN

The weathermen at the two TV stations in town—WET and WILD—were locked in a constant battle to come out with the most accurate forecasts. The WET weatherman's long-term accuracy record was ¾; the WILD weatherman's long-term accuracy record was ⅘.

For the day ahead, WET predicted rain, while WILD predicted sun. Assuming that rain and sun were objectively as likely as one another, what were the odds of rain based on the two stations' forecasts?

73. TWO-WAY ADDITION

Consider the following numbers that appear on a calculator:

Place the numbers into the diagram so that right side up the sum works and upside down it also works. Digits can be repeated and any digit can appear in any spot.

74. THE ICING ON THE CAKE

A square birthday cake measures 12 inches by 12 inches and is six inches high. Three children wish to share the cake equally. But simply dividing the cake into three pieces isn't good enough: the cake has chocolate frosting on all sides except the bottom, and each child wants to have the same amount of chocolate frosting as well as the same amount of cake.

How can the cake be divided into three pieces so that everyone is happy? Since there's a rose made of icing in the center of the top of the cake, and the birthday boy wants the entire rose, no cut may pass through or touch where the rose is.

75. HIGH MATH AT THE 7-11

A customer brought four items to the cashier of the 7-11 convenience store. "That'll be $7.11," the clerk said. At first the customer thought it was a joke. "Ha, ha. Seven-eleven. I get it." But the clerk was serious. "No, really. I multiplied the prices of the four items you gave me, and I came up with $7.11."

"You *multiplied* them?" the customer asked. "You're supposed to *add* them, you know."

"I know," said the clerk, "but it doesn't make any difference. The total is still $7.11."

What were the prices of the four items?

76. WHO AM I?

I am a number with the following properties:
  If I am not a multiple of 4, then I am between 60 and 69.
  If I am a multiple of 3, I am between 50 and 59.
  If I am not a multiple of 6, I am between 70 and 79.
  What number am I?

77. HOME ON THE RANGE

Imagine a pasture that is just big enough to feed 11 sheep for a total of 8 days. It turns out that if we reduce the number of sheep to 10, they would be able to eat for 9 days.

Theoretically, how long could two sheep last?

78. PARTY OF 12

Mr. and Mrs. Green were planning a dinner party for themselves and five other couples. They agreed that no husband and wife should be seated next to one another, so Mrs. Green devised the following seating plan:

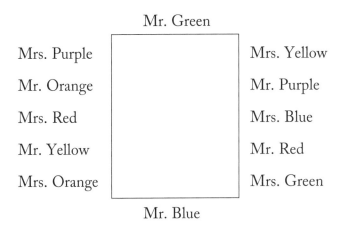

Mr. Green

Mrs. Purple		Mrs. Yellow
Mr. Orange		Mr. Purple
Mrs. Red		Mrs. Blue
Mr. Yellow		Mr. Red
Mrs. Orange		Mrs. Green

Mr. Blue

But Mr. Green insisted that he and his wife sit at opposite ends of the table, so he devised a different plan:

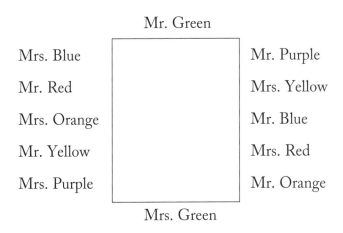

Mr. Green

Mrs. Blue		Mr. Purple
Mr. Red		Mrs. Yellow
Mrs. Orange		Mr. Blue
Mr. Yellow		Mrs. Red
Mrs. Purple		Mr. Orange

Mrs. Green

Mrs. Green would have nothing of it. She was upset that her husband's new seating arrangement didn't follow a man-woman-man-woman alternating pattern all the way around the table, the way hers did.

Can you come up with a seating plan that will (sneakily) satisfy the conditions put forward by Mr. and Mrs. Green?

79. HEADS OR TAILS

If you flip a coin five times, what is the probability that three or more consecutive flips come out the same?

80. AS EASY AS PI

In the diagrams below, which shaded region is bigger?

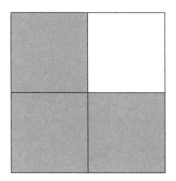 

81. TAKING THE LONG SHOT

Right before the fifth race at Santa Anita, the word got out that one of three horses was going to win the race: Lightning Bolt, a 2–1 shot; Golden Honey, a 3–1 shot; or Matchmaker, a 4–1 shot.

Assuming that this tip is true, how much would you have to bet on each of the three horses to be assured of coming out with a $39 profit?

82. TRICK TRIG QUESTION

When is the equation below true?

$$\frac{\sin(x)}{n} = 6$$

83. MAPPING IT OUT

If you take a map of the United States and draw a line from each city to the next-closest city, what is the maximum number of lines that can emanate from any one city?

84. CLOSE COUNTS IN HORSESHOES

Arrange the numbers 1 through 15 in a horseshoe pattern in such a way that any two consecutive numbers add up to a perfect square.

— — —
—   —
—   —
—   —
—   —
—   —
—   —

85. SEE YOU LATER, ESCALATOR

Dave is in a hurry as he approaches the airport escalator, but Frank is in an even bigger hurry. For every two steps Dave takes, Frank takes three. Altogether, Frank takes 25 steps before dismounting the escalator, while Dave takes 20. Both Dave and Frank step on every step. How many steps does the escalator have showing at any given time?

86. WHITE ON RED

A drawer contains some number of white socks and red socks. It turns out that if you remove two socks from the drawer at random, the chance of getting two white socks is precisely one-third. Given that the drawer has between 4 and 30 socks, how many white and red socks are in the drawer?

87. THE 18-12 OVERTURE

Fill in the remaining seven boxes in the diagram below to form a magic square—one in which the numbers in each row, column, and diagonal add up to the same number. (There is more than one solution.)

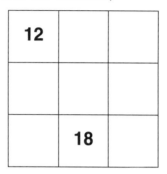

88. SITTING ON THE FENCE

Suppose that a babysitter had a choice of two payment options: 1) Accepting a flat rate of $15 for a night's work, or 2) choosing among six envelopes, two containing $1 bills, two containing $5 bills, one with a $10 bill, and one with a $20 bill.

Note that the second option could leave you with as little as $2 (choosing $1 both times) or as much as $30 (choosing the $10 and the $20). Assuming that you were going to babysit for a long, long time, which is the better payment option?

89. TURNING ON A DIME

Suppose that the diameter of a dime was one-fourth the diameter of a silver dollar. (In real life, the dime is bigger than that, but no matter.) If you placed the dime at the top of the silver dollar and rotated it around the dollar's circumference without slippage—returning to the original position—how many complete revolutions would the dime undergo?

90. RAZOR-SHARP REASONING

Below are the three heads of an electric razor. If each head is one inch in diameter, how far is it all the way around the razor?

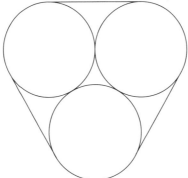

91. FAIR AND SQUARE

Can you find five distinct integers A, B, C, D, and E, each less than 10, so that the equation below is true?

$$A^2 + B^2 + C^2 + D^2 = E^2$$

92. SPECIAL DELIVERIES

Suppose that in the year 2050, Sunnyside Hospital will deliver 200 babies, while Busybee Hospital will deliver 1,000 babies. See if you can guess which hospital has the greater chance of delivering precisely as many boys as girls. (Assume that the probability of having a boy or a girl is the same.)

93. BATTLE STATIONS

A battalion 20 miles long advances 20 miles. During this time, a messenger on a horse travels from the rear of the battalion to the front and immediately turns around, ending up precisely at the rear of the battalion upon the completion of the 20-mile journey. How far has the messenger traveled?

94. SPENDING SPREE

Suppose that one bowl costs more than two plates, three plates cost more than four candlesticks, and three candlesticks cost more than one bowl. If it costs precisely $100 to purchase a plate, bowl, and candlestick, how much does each item cost?

95. A TWO-AND-ONE COUNT

The sum of the first N integers equals 2000. Wait a minute, that's impossible—one of those integers must have been counted twice!

That's the question. If in fact one and only one number was double-counted, which number must that have been?

96. RATIONAL THINKING

If the shaded region below is a square, what is the length of one of its sides?

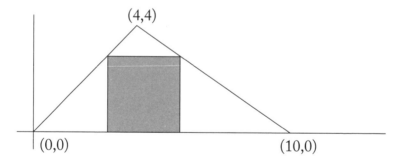

97. ROAMIN' CANDLES

Suppose you have two candles of the same height but of different widths. One takes 4 hours to burn all the way down, while the other takes 7 hours. Assuming both candles burn down at steady rates, how long will it take before one candle is twice as tall as the other?

98. I'M EIGHTEEN, AND I LIKE IT

What is the smallest number that contains precisely 18 proper factors? (A proper factor is a factor other than 1 or the number itself.)

99. MILES TO GO

Sid Jones was driving in his recently purchased sports car when he noticed that the odometer read precisely 12345.6 miles. What is truly amazing is that his trip odometer underneath read precisely 123.4 miles. What is the smallest distance that Jones can drive so that the two odometers have all ten digits between them, but share no digits in common?

100. UPSIDE-DOWN PYTHAGORAS

What is the smallest solution, in whole numbers, to the equation below?

$$\frac{1}{x^2} + \frac{1}{y^2} = \frac{1}{z^2}$$

101. THE ABSENT-MINDED PROFESSOR

A probability professor misplaced his latest set of test papers. In a panic, he decided to give his students random grades, from 0 to 100. The only problem is, two of the students in his class are the Carlson twins, who study together and almost always obtain similar grades. His fear is that if the twins' scores differ by more than 20 points, they will suspect that something is wrong.

The problem boils down to this: If you pick two numbers at random from the interval 0–100, what is the probability that those two numbers will be within 20 of one another? (Assume that the scores do not have to be whole numbers.)

Answers

1. WHEN IN ROME

Because X × X < 10, it follows that X = 1, 2, or 3. However, X = 1 is obviously impossible, because then XLIV would have to equal CDXL. It can also be shown that X = 3 doesn't work: L would then have to equal 1 or 2, and either choice produces a contradiction. If L = 1 then V = 7, but then I would also have to equal 7 to produce the X in CDXL. Similarly, if L = 2 then V = 4 but I would also have to be 4.

It follows that X = 2, and it turns out that there are two solutions: 2864 × 2 = 5728 and 2814 × 2 = 5628.

2. FIND THE SHORTCUT

The answer is 53. If 5^3 = 125, then 50^3 = 125,000. Similarly, 60^3 = 216,000. Note that 148,877 is in between 125,000 and 216,000, so if 148,877 is the cube of some whole number, that number must be between 50 and 60. But the only way the cube of a number can end in 7 is if the original number ends in 3. (In particular, 3^3 equals 27. The general pattern can be checked by trial and error: The cube of a number ending in 1 also ends in 1, the cube of a number ending in 2 ends in 8, and so on.) Therefore the cube root of 148,877 must be 53.

3. PAGE BOY

The sum must have been 412, because 412 = 48 + 49 + 50 + 51 + 52 + 53 + 54 + 55.

512, on the other hand, is a power of two, and no power of two can be expressed as the sum of consecutive integers. The proof of this follows:

Note that the sum of the first n integers is given by $n(n+1)/2$. The sum of consecutive integers beginning with m + 1 and ending with n is given by $n(n+1)/2 - m(m+1)/2 = ((n^2 - m^2) + (n - m))/2$.

But $n^2 - m^2 = (n + m)(n - m)$, so this expression becomes $(n + m + 1)(n - m)/2$. Now note that $(n + m + 1)$ and $(n - m)$ have different parities—meaning that if one is even, the other must be odd. Therefore their product cannot be a power of two, which completes the proof.

4. CIRCULAR LOGIC

Note that the distances XY and YZ must each be less than the diameter of the circle. Also, segments XC and CZ must sum to precisely the diameter of the circle, because each one of these two segments is a radius.

Therefore, the sum of all four of these segments must be less than three times the diameter of the circle. But the circumference equals π times the diameter, and π is greater than 3, so the family member who walks along the straight paths arrives home first.

5. ACROSS TO BEAR

The key to finding the number of the final across entry is using the fact that all crossword puzzles are rotationally symmetric with respect to the center square. (If there is a black square in the first row of the fifth column, there must be a black square in the bottom row of the 11th column, as there is below.) Therefore the whole puzzle must look like this, with the final across entry being number 67.

1	2	3	4		5	6	7	8		9	10	11	12	13
14					15					16				
17					18					19				
20					21			22						
23			24					25						
			26				27			28	29	30		
31	32	33			34	35				36				
37					38			39	40					
41				42			43							
44			45	46				47						
			48			49	50			51	52	53		
54	55	56			57			58						
59					60			61						
62					63			64						
65					66			67						

6. NO CALCULATORS, PLEASE

First note that $(\frac{1}{2}) \times (\frac{2}{3}) \times (\frac{3}{4}) \times (\frac{4}{5}) \times (\frac{5}{6})... \times (\frac{97}{98}) \times (\frac{98}{99}) \times (\frac{99}{100}) = \frac{1}{100}$, which is readily seen because everything other than the initial 1 and the trailing 100 will cancel out.

Therefore the following two numbers multiply to $\frac{1}{100}$:

$(\frac{1}{2}) \times (\frac{3}{4}) \times (\frac{5}{6}) \times ... (\frac{97}{98}) \times (\frac{99}{100})$ and
$(\frac{2}{3}) \times (\frac{4}{5}) \times (\frac{6}{7}) \times ... (\frac{98}{99}) \times 1$

The top number is the one we are interested in. Clearly it is less than the bottom number, because it is less on a term-by-term basis—$\frac{1}{2} < \frac{2}{3}$, $\frac{3}{4} < \frac{4}{5}$, etc. The top number must therefore be less than the square root of $\frac{1}{100}$, or $\frac{1}{10}$.

7. BORN UNDER A BAD SIGN

The answer is E), Sue's 50th birthday.

Clearly Sue's birthday will always be on the 13th, so the only question is when it will be on a Friday. The years that produce a Friday birthday (the same logic holds for any day of the week) follow a cycle of 6, 11, 6, and 5 years: Note that 6 + 11 + 6 + 5 = 28, and after 28 years the cycle repeats itself. (There are 4 years between leap years and 7 days in the week, so calendars repeat themselves every 4 × 7, or 28, years.)

To see how the 6, 11, 6, 5 sequence comes up, suppose you were born on a Friday in March of a leap year. The following year your birthday will fall on a Saturday, because 365 has a remainder of 1 upon division by 7. The year after that your birthday will be on a Sunday, then Monday, Wednesday (the key step—skipping a day because this year in the sequence is a leap year), then Thursday and, finally, Friday. In other words, in 6 years your birthday will again be on a Friday, and you'll be precisely between two leap years. Continuing, the next Friday birthday will occur in 11 years, because the 3 leap years during this period will push your birthday up by a total of 11 + 3 = 14 days, or precisely two weeks. The remainder of the sequence is found the same way, at which point you're back to a leap year, and you start all over again.

We don't know when in the cycle Sue was born, but we can still answer the problem by looking at the sequence 6, 11, 6, 5, 6, 11, 6, 5, 6, 11, 6, 5, 6, 11, 6, 5, 6, 11, 6, 5, and trying to find consecutive entries that sum up to a multiple of 10—10, 20, 30, 40, 50, or 60.

The only one of these multiples of 10 that works is 50, which equals 5 + 6 + 11 + 6 + 5 + 6 + 11. Specifically, if Sue was born on Friday the 13th during a year prior to a leap year (such as 1987), her 50th birthday will also fall on Friday the 13th. None of the other birthdays listed will ever fall on a Friday, no matter what year she was born in.

8. CROWD O' THREES

Here are some representations of the numbers 1 through 20, using precisely three threes:

$1 = 3^{3-3}$

$2 = \sqrt{3 + \frac{3}{3}}$

$3 = 3 + 3 - 3$

$4 = 3 + \frac{3}{3}$

$5 = 3! - \frac{3}{3}$

$6 = \sqrt{33 + 3}$

$7 = 3! + \frac{3}{3}$

$8 = \left(\sqrt[3]{3}\right)^3$

$9 = 3 + 3 + 3$

$10 = 3(3 + .\overline{3})$

$11 = \frac{33}{3}$

$12 = 3 \times 3 + 3$

$13 = \frac{3}{.3} + 3$

$14 = \frac{3}{.3} - 3!$

$15 = 3! + 3! + 3$

$16 = \frac{3}{.3} + 3!$

$17 = \frac{3}{.3} - 3$

$18 = 3 \times (3 + 3)$

$19 = 3 \times (3! + .\overline{3})$

$20 = \frac{(3 + 3)}{.3}$

Note that the bar used in the solutions for 10 and 19 indicates a repeating decimal: $.\overline{3} = \frac{1}{3}$.

9. TENNIS, ANYONE?

Suppose the records of the players are as follows:

Player	Record
A	7–0
B	6–1
C	5–2
D	4–3
E	3–4
F	2–5
G	1–6
H	0–7

In this case, the total amount paid out would be $7^2 + 6^2 + 5^2 + 4^2 + 3^2 + 2^2 + 1^2 = 140$. In other words, 140 is the sum of the first seven squares. Now, it is not necessarily the case that each player wins a different number of sets, but note that any change to the above table will reduce the total cash outlay. For example, if you changed player A's record to 6–1 and player F's record to 3–4 (keeping total numbers of wins and losses the same), you'd now have only 132 dollars in total prizes.

But what if the results were as follows?

Player	Record
A	7–0
B	6–1
C	6–1
D	4–3
E	3–4
F	1–6
G	1–6
H	0–7

Now you'd have a total of 49 + 36 + 36 + 16 + 9 + 1 + 1, or 148 dollars to pay out. But the above pattern is impossible! If A beat B and C—accounting for the only loss suffered by those two players—then what happened when B played C? Similarly, F and G must have both beaten H, but there is no room for them to have played each other. This illustrates why the pattern above, with a payout of $140, must be maximal. Since $17.50 from each of the 8 participants yields $140, the entry fees are always enough to pay for the prizes.

10. ONE, TWO, THREE

There are four solutions, as follows:

1	2	3	4
192	219	273	327
384	438	546	654
576	657	819	981

11. DON'T MAKE MY BROWN EYES BLUE

The surprising answer is that everyone with blue eyes vacated the island the morning after the tenth day following the decree.

To see why this is so, suppose that only one person on the island had blue eyes. (Recall that at least one person had blue eyes. This condition may have seemed unimportant at the time, but it was essential to get the proof rolling. This is true of all proofs based on "mathematical induction.") Anyway, after one day, that person, having seen everyone else on the island, would have been able to conclude that he or she had blue eyes, and would therefore leave.

Similarly, if two people had blue eyes, they would see each other on the first day. Then, on the second day, each would see the other, and each would then realize that if that other person was the only one on the island with blue eyes, that other person would have left after the first day. Accordingly, the two would independently realize that they each had blue eyes, and would leave after the second day. And so on, and so on.

12. THE STAMP COLLECTION

We know right away that the number of stamps in the collection is divisible by 35 since the number of stamps can be divided evenly by 5 (a fifth of the stamps are in Book 1) and 7 (some number of sevenths are in Book 2) and because 5 and 7 have no common factor. Suppose there are $x/7$ of the stamps in the second book. Together, the first and second books contain $1/5 + x/7 = {(7 + 5x)}/35$ of the collection. The third book therefore contains ${(28 - 5x)}/35$ of the collection. If there are C stamps in all, then $35 \times 303 = C \times (28 - 5x)$. But 35 divides into C, so $(28 - 5x)$, which is a positive integer, must equal one of the factors of 303: 1, 3, 101 or 303. Try each of these cases. The only one that leads to a positive integer less than 7 is $(28 - 5x) = 3$, giving $x = 5$. This yields that 303 stamps amounts to $3/35$ of the collection, so the entire collection equals $35 \times 101 = 3{,}535$ stamps.

13. TOO CLOSE TO CALL

Take the thirtieth power of both numbers, so as to remove the radical signs altogether. We get the following:

$$\sqrt[10]{10}^{\,30} = 10^3 = 1000$$
$$\sqrt[3]{2}^{\,30} = 2^{10} = 1024$$

Therefore the cube root of 2 is slightly larger. (Note that the "K" in the computer memory context is actually 1,024—a power of two, befitting the binary code—but its proximity to 1,000 causes it to be used interchangeably, as in Y2K, etc.)

14. THE LONG STRING

The answer is an emphatic yes. The 1,000,000-term sequence 1,000,001! + 2, 1,000,001! + 3, all the way up to 1,000,001! + 1,000,001 consists entirely of composite numbers, because 1,000,001! + K is always evenly divisible by K for any K in this range. More generally, no matter how big the number N is, it is always possible to find N consecutive composite numbers.

15. FIRST-CLASS LETTERS

The only number between 1 and 5,000 that is alone in its letter class is 3,000. THREE THOUSAND requires 13 letters to write out, and is the *only* number within that range to require precisely 13 letters.

Although we won't provide an exhaustive proof, you would begin by noting duplications among the first nine digits, as in ONE-TWO-SIX, THREE-SEVEN-EIGHT,

FOUR-FIVE-NINE. Then we would find duplications among TWENTY and THIRTY, and so on, which would rule out solutions with two digits. The three-digit case is similar. In fact, there would be duplications in the four-digit category as well, except that we restrict ourselves to the 1–5,000 range: Note that 7,000 is also in letter class 13.

16. MOST VALUABLE PUZZLE

1) Seven first-place votes would do the trick. Whoever won seven first-place votes would have a minimum of $(7 \times 3) + 3 = 24$ points. The best anyone else could do would be to win 3 first-place votes and 7 second-place votes, for a total of $(3 \times 3) + (7 \times 2) = 23$ points.

2) In theory, the MVP award could be won by someone with only one first-place vote, as long as that person received a second-place vote on each of the remaining 9 ballots. The voting would have to look like this:

Player	1st	2nd	3rd	Points
A	1	9	0	$(1 \times 3) + (9 \times 2) + (0 \times 3) = 21$
B	5	0	5	$(5 \times 3) + (0 \times 2) + (5 \times 1) = 20$
C	4	1	5	$(4 \times 3) + (1 \times 2) + (5 \times 1) = 19$

17. PERFORMANCE ANXIETY

The basis of the misleading arithmetic is the confusion between a percentage increase and a percentage-point increase. The actual performance of the firm's selections, in percent, was a number 32 percent bigger than 28, namely, 28×1.32, or about 37 percent. Not too shabby, but a far cry from a return of 60 percent, which would be a 32 *percentage-point* improvement over the S&P 500.

18. DOUBLE TROUBLE

$58 \times 3 = 174 = 29 \times 6$

19. CEREAL SERIAL

The answer is 8⅓ boxes. To see why, note that one of the prizes comes from the first box we buy. The likelihood of getting a new prize from the next box equals ¾; on average, therefore, we would need to buy ⁴⁄₃ boxes to get a new prize. Proceeding in this same fashion, the third new prize would require an additional $\frac{1}{(½)} = 2$ boxes. The fourth would require, on average, an additional 4 boxes. In total, the average number of boxes equals $1 + ⁴⁄₃ + 2 + 4 = 8⅓$ boxes.

In real life, of course, you can't buy ⅓ of a box, but that is still the average number of boxes you'd have to purchase.

20. PAIRING OFF

The next pair (x,y) satisfying $x^2 - 2y^2 = 1$ is (17,12), and it is easy to check that $17^2 - 2(12^2) = 289 - 2(144) = 1$.

Believe it or not, there is a systematic (though hardly obvious) way of generating all solutions to the equation. If you consider the quantity $(3 + 2\sqrt{2})$, you will discover that $(3 + 2\sqrt{2})^2 = 17 + 12\sqrt{2}$. If you continued taking successive powers and examining the coefficients, you would discover that the next ordered pair higher than (17,12) is (99,70), and so on.

21. A SQUARE DEAL

The proper dimensions are as shown. There is no solution if the smallest square is only one mile on each side. There is, however, a solution when the smallest square has a side of two. The plots of the second and third child each measure 96 square miles, and the total area of the father's plot was 196 square miles.

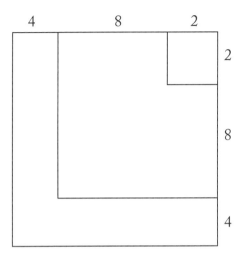

22. SURVIVAL OF THE SPLITTEST

This is one of those problems where it's just as easy to assume that the probability of a successful split is a variable p, rather than a specific number such as ¾. So let p be the probability of a successful split, and P be the probability that the amoeba chain will go on forever.

Look at the second generation of amoebas. With probability p, there are two amoebas in that generation. The probability that these two will generate an infinite chain is $1 - (1 - P)^2$, because $(1 - P)^2$ is the probability that *neither* will do so. Therefore, $P = p(1 - (1 - P)^2)$, because both sides of the equation represent the probability of long-term survival.

Simplifying, we get the equation $pP^2 + (1 - 2p)P = 0$, or $P(pP + (1 - 2p)) = 0$. We assume that P does not equal 0, so that $pP + (1 - 2p) = 0$, or $P = {}^{(2p-1)}\!/_p$.

Setting p = ¾, we see that $P = (2(¾) - 1) \div (¾) = ⅔$.

23. SHUTTING THE EYE

The radius of the big circle is five times the radius of the smallest circle. To see why, consider the diagram below:

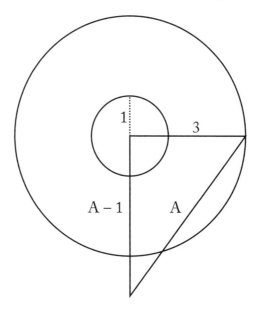

Let's say that the radius of the small circle is 1, and so the radius of the mid-sized circle is 3. If A is the radius of the big circle, then we see that $3^2 + (A - 1)^2 = A^2$. This equation simplifies to $9 - 2A + 1 = 0$, or $A = 5$. (You could also see that the indicated equation yields the Pythagorean relationship for the famous 3–4–5 right triangle. However you slice it, $A = 5$, so the radius of the largest circle is five times that of the smallest one.)

24. WHAT'S IN A NAME?

Of the three cryptarithms, only the middle one has a solution. Here's why.

$$
\begin{array}{r}
\text{M A R Y} \\
+ \ \text{T Y L E R} \\
\hline
\text{M O O R E}
\end{array}
$$

Sorry, Mary, but your equation cannot be solved. Look at the second column. In order for R + E to equal R, E must be either 0 or 9: If E = 0, that means there is no carrying from the first column; if E = 9, there must be a 1 carried over from the first column. Unfortunately, if E = 0 in the first column, there would be a 1 carried over, while if E = 9, there would be nothing carried over—precisely the opposite of our requirements for column two! The addition is therefore impossible.

$$
\begin{array}{r}
\text{J O Y C E} \\
+ \ \text{C A R O L} \\
\hline
\text{O A T E S}
\end{array}
$$

There are many solutions to this one. One solution is given by the following:

$$
\begin{array}{r}
5 \ 9 \ 8 \ 3 \ 2 \\
+ \ 3 \ 7 \ 1 \ 9 \ 4 \\
\hline
9 \ 7 \ 0 \ 2 \ 6
\end{array}
$$

Observe that the solution is not unique. In particular, you can create another solution by simply interchanging the 1 and the 8 in the middle column.

$$\begin{array}{r} \text{J A M E S} \\ - \quad \underline{\text{E A R L}} \\ \text{J O N E S} \end{array}$$

This one has no solution, and to see why you don't need to look beyond the trailing letters. It is impossible for ES – RL to equal ES unless both R and L are zero. But two different letters cannot be given the same number, so no solution is possible.

25. THUMBS DOWN

The key is to notice that a must be even. But it can't be 4 or 6, because the product would exceed three digits. So a must equal 2. With a little trial and error, the other digits fall into place. The only solution is 275 × 2 = 572 in base 8.

26. TICKET TO RIDE

If F = the number of former stations and N = the number of new ones, we must have $2FN + N(N - 1) = 52$. This equation becomes $N^2 + (2F - 1)N - 52 = 0$. We know that -52 is the product of the roots of this equation, and since the only possible factorizations of 52 are 1×52, 2×26, and 4×13, the sum of the roots must equal 51, -51, 24, -24, 9, or -9. But this sum must equal $2F - 1$, with F a positive integer, so the only possibility is $2F - 1 = 9$, in which case F = 5. The equation now factors as $(N + 13)(N - 4) = 0$, and the positive root is N = 4. (The possibility $2F - 1 = 51$ yields F = 26 and N = 1, but the problem stated that several new stations—plural —were added, implying that N > 1.) So there were 5 stations originally, and 4 were added.

27. GOING OFF ON A TANGENT

Any two tangents to a circle must have equal length, so we can segment the lengths of the diagram as follows:

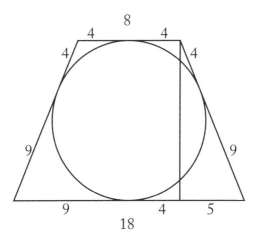

The "5" appears because the length of that segment is the difference of the lengths of two other segments, now known to be 9 and 4, respectively. But the diameter of the circle can now be seen to be a leg of a right triangle with hypotenuse 13 (= 9 + 4) and other leg 5, so by the Pythagorean theorem the diameter must be 12.

28. SURPRISE ENDING

The expression $1^n + 2^n + 3^n + 4^n$ is divisible by 5 unless n is divisible by 4. The most straightforward method of proving this result uses modular arithmetic. We start with the following table:

n	1^n	2^n	3^n	4^n
1	1	2	3	4
2	1	4	9	16
3	1	8	27	64
4	1	16	81	256
5	1	32	243	1024

Now we take the remainders upon division by 5, and add those remainders up:

n	1^n	2^n	3^n	4^n	$1^n + 2^n + 3^n + 4^n$
1	1	2	3	4	10
2	1	4	4	1	10
3	1	3	2	4	10
4	1	1	1	1	4
5	1	2	3	4	10

Note that n = 4 produces the same results as n = 0 would have, and then the pattern repeats itself. The result is surprising in that a question about divisibility by one number (5) turns out to be related to divisibility by a relatively prime number (4).

29. THANKSGIVING FEAST

Let T denote the set of pilgrims that had turkey, C the set that had corn, P the pumpkin pie eaters, and M the mince pie eaters. Similarly, let T', C', P', and M' denote the set of people who did not have the indicated dishes—i.e., the complements of the respective sets. Then T' consisted of 10 percent of the pilgrim population, C' 20 percent, P' 30 percent and M' 40 percent. Because no one had all four dishes, the union of these sets is the entire pilgrim population. Also, since 10 + 20 + 30 + 40 equals precisely 100, these four sets must be disjoint. In particular, that means that P' and M' have no members in common, so everyone must have had one of the two desserts.

30. ALL ABOUT PYTHAGORAS

1) The only numbers that cannot be part of a Pythagorean triple are 1 and 2: In other words, there are no two perfect squares that differ by either 1 or 4. To see why any other number n must be part of a triple, first suppose that n is odd. Then n can be represented as the shorter leg as in the diagram below. Note that the longer leg is always one less than the hypotenuse, as in (3,4,5), (5,12,13), and (7,24,25).

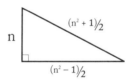

Now for the case where n is even. If n = 4, we already have a solution—namely, the (3,4,5) right triangle. If n is greater than or equal to 6, let n = m × k, with m odd. We can form a Pythagorean triple using m, as above. Then we simply multiply all the sides by k. (The final case, where n is a power of 2, is similar.)

2) 15 is the smallest such number. It is the hypotenuse of the 9–12–15 triangle (obtained by multiplying the 3–4–5 triangle by 3 on each side); it is the smaller leg of the 15–36–39 triangle (obtained by multiplying the 5–12–13 triangle by 3); and it is the larger leg of the 8–15–17 right triangle.

31. A DECADENT DECADE

The answer is the 1870s, because 1871, 1873, 1877, and 1879 are all prime numbers.

To prove this result, note first the obvious point that only those years ending in 1, 3, 7, or 9 are candidates for primality. Also note that each of these digits is congruent to either 0 or 1 (mod 3)—in other words, the remainder of the sum upon division by 3 must be 1. Therefore, the sum of the first three digits of the decade must be congruent to 1 (mod 3), because otherwise one of the four candidates in that decade would be divisible by 3.

From 1776 to the year 2000, there have been eight decades whose first three digits add up to be 1 (mod 3), as follows: 1780s, 1810s, 1840s, 1870s, 1900s, 1930s, 1960s, 1990s.

We can rule out all but one of these decades by simple trial and error, which produces the following results:

$1781 = 13 \times 137$
$1813 = 7 \times 259$
$1841 = 7 \times 263$
$1903 = 11 \times 173$
$1939 = 7 \times 277$
$1969 = 11 \times 179$
$1991 = 11 \times 181$

That leaves the 1870s as the only real candidate, and indeed it turns out that 1871, 1873, 1877, and 1879 are all prime, which admittedly can be deduced only laboriously or by computer. But if you take our word that there was such a decade, the above shows that the 1870s must be it.

32. COMPOSITE SKETCH

The first all-composite decade consisted of the years 200 through 209. Note that 201 = 3 × 67, 203 = 7 × 29, 207 = 9 × 23, and 209 = 11 × 19; the other years in the decade are composite since they are even or divisible by 5.

33. FIVE SQUARES TO TWO

The diagram below shows where the cuts should be made. Note that the two pieces that must be moved to form the 1 × 2 rectangle arc of the same size and shape.

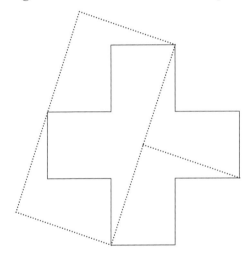

34. EQUALITY, FRATERNITY

1) $6^14^5 - 1 = 6143$
2) $-127 = 1 - 2^7$
Other solutions are possible.

35. BLUE ON BLUE

The smallest possible value of N is 10. Specifically, if there is one red ball and nine blue balls, the probability that the first selected ball is blue equals $\frac{9}{10}$; the probability that the second ball is blue equals $\frac{8}{9}$, and so on. The probability that all five balls are blue equals $(\frac{9}{10})(\frac{8}{9})(\frac{7}{8})(\frac{6}{7})(\frac{5}{6}) = \frac{5}{10} = \frac{1}{2}$.

This same result can be proved without this sort of computation, but note that the product is trivial to calculate because all of the factors cancel out except for 5 and 10. In general, if there are n balls selected, the smallest value of N that produces a probability of $\frac{1}{2}$ for all n balls to be blue equals 2n.

36. STICK FIGURES

A) $1\,1 = -\,1\,1 - 3\,3 + 5\,5$

B) $\overline{1\,1} = 1\,1\,3\,\overline{)\,3\,5\,5}$

Yes, that's π on the left hand side of the second equation. Remarkably, $\frac{355}{133}$ equals 3.14159292035..., whose first six decimal places match those of π (3.1415926535...).

37. HERALDING LOYD

The beauty of the puzzle is that you don't need to know anything about the boats' relative speeds to figure out the width of the river (although you can certainly deduce the relative speeds after obtaining the answer).

When the boats first meet, the total distance they have traveled equals the width of the river. By the time they meet again, the total distance traveled equals *three* times the width of the river. (Draw a diagram to convince yourself.) The boats are each traveling at a constant speed, so they each will have traveled three times as far by the second meeting as the distance they'd traveled by the first time they met. Because the boat starting in New York had traveled 720 yards at the first meeting, it must have traveled 2,160 yards at the time of the second meeting. But this distance is 400 yards from the other shore, so the width of the river equals 2,160 − 400 = 1,760 yards. Conveniently, this is exactly one mile.

38. LOTS OF CONFUSION

If N is the number of lots and I is the original price per lot, we can obtain the following equations (each of the three expressions represents net profit):

$$18N − 243 = 6I = N(18 − I)$$

From the first and third equations we get $I = {}^{243}/_N$, and substituting into the second equation and combining with the first yields the equation $18N^2 − 243N − (6 × 243) = 0$. Dividing by 9 produces $2N^2 − 27N − (6 × 27) = 0$. This factors into $(2N + 9)(N − 18) = 0$, so we see that N = 18.

39. OH, HENRY!

The answer is 21 square miles. The puzzle can be solved in an unusual manner by noticing some interesting facts about the areas given in the question: $388 = 8^2 + 18^2$ and $153 = 3^2 + 12^2$ and $61 = 5^2 + 6^2$. Using this information, you can construct the marvelous diagram below. Here, line segment A stands for one side of region A, etc. But the area of the thin triangular region is simply the area of the big triangle—$\frac{1}{2} × 8 × 18$—minus the areas of the two smaller triangles and rectangle

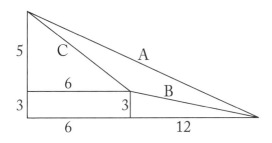

(½ × 5 × 6, ½ × 3 × 12, and 3 × 6, respectively). This yields
72 − (15 + 18 + 18) = 72 − 51 = 21 square miles.

40. CATCH-22

The numbers whose factorials end with precisely 22 zeroes
are 95, 96, 97, 98, and 99.

To see why, note that every multiple of 10 between 10 and 90 contributes one zero to the factorial, for nine altogether. And every combination of 2 and 5—of which there are 10 between 1 and 95, inclusive—contributes another zero, so the total is up to 19. But 25, 50, and 75 each have *two* factors of 5, each of which produces another zero when coupled with any other even number. That makes three more, for a grand total of 22. That total is shared by 95!, 96!, 97!, 98!, and 99!. Note that multiplying 99! by 100 immediately adds *two* more zeroes, so there is no number whose factorial has precisely 23 zeroes at the end.

41. WALKING THE BLANK

The probability is one. No matter how you fill in the digits, the resulting 28-digit number will be divisible by 396.

To see why this works, note that 396 = 4 × 9 × 11. At this point we simply need to know the proper divisibility tests for 4, 9, and 11, which are stated below.

4: The last two digits form a number that is divisible by 4

9: The sum of the digits must be divisible by 9

11: The sum of the digits in odd positions minus the sum of the digits in even positions must be divisible by 11.

Note that the number ends in 76, which is divisible by 4, and therefore so is the entire number. And the sum of the existing 18 digits equals 90, which is divisible by 9, as is the sum of 0 through 9, so the whole number is divisible by 9. It only remains to check for divisibility by 11. The sum of the "odd" digits is 5 + 3 + 3 + 8 + 2 + 9 + 6 + 5 + 8 + 2 + 3 + 9 + 3 + 7 = 73, while the sum of the "even" digits is 8 + 3 + 0 + 6 + (sum of 1 through 9) = 8 + 3 + 0 + 6 + 45 = 62. The difference of 73 and 62 is 11, which is of course divisible by 11. The entire number must always be divisible by 396, no matter where the digits 0 through 9 are placed. This puzzle was originally created by Leo Moser—as an April Fool's prank!

42. BREAKING THE HEX

The larger hexagon is precisely three times as big as the smaller hexagon. To see why, create the following diagram, where each of the six equilateral triangles in the original diagram is essentially flipped on its base, forming six more equilateral triangles in the middle. But each of the other six (isosceles) triangles has the same area as each of the equilateral triangles! (They have the same bases, and the altitude of any of the equilateral triangles, as dropped from one of the vertices of the large hexagon, is also an altitude for the isosceles triangle.) All in all, the large hexagon contains 18 triangles of the same area, and the small hexagon contains 6 such triangles, so the large hexagon is three times as big.

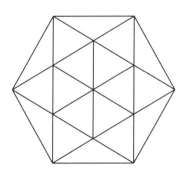

43. DOMINO THEORY

The probability is $^{17}/_{48}$.

The key observation is that in order for a continuous chain of dominoes to be made in the manner described, the leftmost and rightmost spots must be the same. (If different, there would be no way for either to appear an even number of times, which is essential.) Now we make the teensy-weensy (but valid) assumption that for any pair of dominoes that share a spot, we can always construct a continuous chain with that spot at either end.

With that assumption under our belts, here is the complete set of ordered pairs that 1) contain at least one 5 and one 6, and 2) contain a pair of repeated spots to serve as the endpoints of the chain:

5–0 & 6–0	5–6 & 6–0	5–0 & 6–5
5–1 & 6–1	5–6 & 6–1	5–1 & 6–5
5–2 & 6–2	5–6 & 6–2	5–2 & 6–5
5–3 & 6–3	5–6 & 6–3	5–3 & 6–5
5–4 & 6–4	5–6 & 6–4	5–4 & 6–5
5–5 & 6–5		
5–6 & 6–6		

The total number of ordered pairs containing at least one 5 and at least one 6 equals 48 (7 × 7 minus the 5–6 domino, which can't be repeated), so the desired probability must be $^{17}/_{48}$.

44. THE WAYWARD THREE

The number is 3,529,411,764,705,882, which when multiplied by ³⁄₂ produces 5,294,117,647,058,823.

To see how to get these numbers, we'll call them A and B, respectively. The first step in determining A is to see that its last digit must be 2. That's because $^{3A}/2$ ends in 3, so $^{A}/2$ must end in 1.

If A ends in 2, B must end in 23, because B is formed from A by putting the 3 at the end. But B = 3 × ($^{A}/2$), so $^{A}/2$ must end in 41 (4 is the only digit whose product with 3 ends in 2). If $^{A}/2$ ends in 41, A ends in 82.

We keep going in this fashion, working right to left, until we (finally) get to a point where we encounter a 3, at which point the successive divisions end. That doesn't happen until the 16th digit!

45. OH, REALLY?

Let $x = \sqrt[3]{2 + \sqrt{5}} + \sqrt[3]{2 - \sqrt{5}}$.

Then $x^3 = 4 + 3\left(\sqrt[3]{\sqrt{5} - 2} + \sqrt[3]{-2 - \sqrt{5}}\right)$, so x satisfies $x^3 + 3x - 4 = 0$. But this equation has only one real root, namely x = 1. [Note that $x^3 + 3x - 4 = (x - 1)(x^2 + x + 4)$, and there are no real solutions to the equation $x^2 + x + 4 = 0$.]

46. DOWN TO THE WIRE

The probability of a four-game sweep by either team is ½ × ½ × ½ × ½ = ¹⁄₁₆, so overall the probability of a four-game Series is ⅛ (because either team could win). The probability of a five-game Series—according to the information supplied in the puzzle—is ¼. Therefore the likelihood of the Series' going either six or seven games is 1 − (⅛ + ¼) = ⅝. But the probability of a six-game Series must equal the probability of a seven-game Series, simply because once game 6 is reached, each team has a 50–50 shot at winning it! Therefore the probability of a seven-game Series is ⁵⁄₁₆.

47. EASY AS A, B, C

The numbers are 2, 3, and 6: $\frac{1}{2} + \frac{1}{3} + \frac{1}{6} = 1$. (This is the only solution with distinct integers. The other solutions are $\frac{1}{2} + \frac{1}{4} + \frac{1}{4} = 1$, and $\frac{1}{3} + \frac{1}{3} + \frac{1}{3} = 1$.)

48. THE BEANPOT RALLY

There are 48 possible combinations for the Beanpot tournament. To see why, note that any one school—say, BU—can have three possible opponents in the first round. After the match-ups are set, each game can end in one of two ways. As there are four games in all (including the third-place playoff), the total number of possibilities equals $3 \times 2 \times 2 \times 2 \times 2 = 48$. (Note that flipping the brackets, or flipping the schools within any one bracket, does not actually add to the possible outcomes.)

49. 3, 4, 6, HIKE!

Let x = the total distance traveled, and y = the uphill (or downhill) distance. Because time = distance ÷ speed, the total time traveled is given by $2(((\frac{x}{2}) - y)/4) + \frac{y}{3} + \frac{y}{6}$, the factor of 2 reflecting the fact that the level portion of the trip was hiked twice.

Therefore $2(((\frac{x}{2}) - y)/4) + \frac{y}{3} + \frac{y}{6} = 5$, which looks like one equation with two unknowns (which would be unsolvable), but in fact the y terms all cancel out, leaving $\frac{x}{4} = 5$, or $x = 20$. The length of the trip was thus 20 miles.

50. EVEN STEVEN

$79 + 5 + \frac{1}{3} = 84 + \frac{2}{6} + 0$

51. SECOND MOST VALUABLE PUZZLE

Player C must have received the second-place vote from the *Herald-Tribune*. To see why, note that there were a total of 39 points awarded, so the total of x, y, and z (the point awards for 1st, 2nd, and 3rd, respectively) must equal 13. The only sets of three distinct integers that add up to 13 are as follows:

x	y	z
10	2	1
9	3	1
8	3	2
8	4	1
7	5	1
6	5	2
6	4	3

But (8,4,1) is the only triplet that can possibly yield a score of 20 points. The point distribution of the balloting must have looked like this:

Player	Herald Tribune	Daily News	Other	Total
A	8	4	8	20
B	1	8	1	10
C	4	1	4	9

Note that Player C garnered both of the second-place votes that didn't go to Player A. In particular, C received the second-place vote of the *Herald-Tribune*.

52. TOP SCORE

The product is maximized when the numbers are 3, 3, 3, 3, 3, 3, and 2. The maximum product is therefore 1,458.

53. A BRIDGE TOO FAR

The answer is six points.

In order for a pair scoring 420 to have received 2½ points, the eight hands must have produced the following scores: 420, 420, 420, 420, 420, 420, 450, 450. (Remember, everyone playing the hand scored either 420 or 450, so the only way to achieve 2½ points would be to tie five other pairs.)

If the pair in question had scored 450 instead, there would be a total of five 420s and three 450s. A pair scoring 450 would get one point for each of the 420s and ½ a point for each of the other two 450s, for a total of six points.

54. SEVEN-POINT LANDING

The diagram below does the trick, and it's not as arbitrary as it may appear. Construct a rhombus with one-inch sides having vertices A, B, C, and D. The distance from A to C should also be one inch. The rhombus is essentially made of two equilateral triangles glued together. Now rotate the rhombus to the right (keeping it fixed at point B) so that point X is one inch away from point D. The seven vertices of the rhombi are the answer.

Check it out. It works!

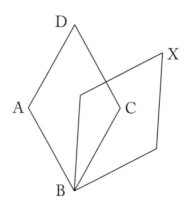

55. 'ROUND GOES THE GOSSIP

The minimum number of phone calls required before everyone knows everything is 8.

Label the busybodies A, B, C, D, E, and F, and divide the six into two groups: A, B, and C are in Group 1; D, E, and F are in Group 2. One sequence of calls that does the job is as follows:

1) A–B
2) A–C (A and C now know all of Group 1)
3) D–E
4) D–F (D and F now know all of Group 2)
5) A–D (A and D now know everything)
6) C–F (C and F now know everything)
7) B–A (B knows everything)
8) E–F (E knows everything)

Note that in step 7, any one of A, C, D, and F could call B, and similarly for step 8. It turns out that 8 is the minimum number of phone calls necessary. In general, if you had n busybodies, the minimum number of calls required before everyone knows everything equals 2n − 4.

56. WHAT'S IN A NAME, PART TWO

$$
\begin{array}{r}
4\ 8\ 6\ 1\ 2\ 8 \\
\times \qquad\qquad 2 \\
\hline
9\ 7\ 2\ 2\ 5\ 6
\end{array}
$$

57. TWO SQUARES ARE BETTER THAN ONE

The first issue to settle is the size of the smaller squares. There are a total of 65 squares to work with. Because $65 = 49 + 16 = 7^2 + 4^2$ is the only other representation of 65 as the sum of two squares, the lengths of the smaller squares must be 7 and 4. By using a stairstep method, we come up with the decomposition below:

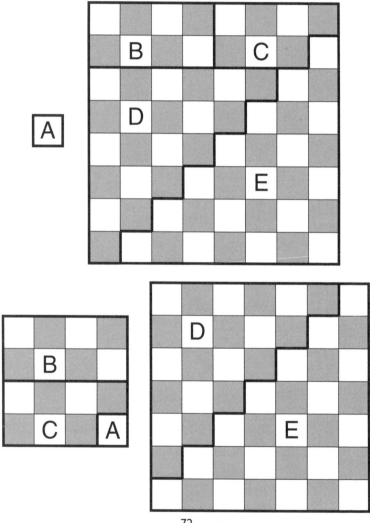

58. HITTER'S DUEL

Going 9 for 12 is equivalent to going 7 for 8 and then going 2 for 4. Assuming the two men had a comparable number of at bats before the games, their averages would still be close—perhaps identical—after both had gone 7 for 8. And as long as their averages were well below .500, Jackson's extra 2-for-4 increment could be enough to raise his average above Cobb's.

59. VISIBLE AND DIVISIBLE

Note that $7 \times 8 \times 9 = 504$. Upon dividing 789,000 by 504, you get a remainder of 240. Therefore, if you add $504 - 240 = 264$ to 789,000, you get the desired number: 789,264. This answer is unique because the only other number of the form 789,XYZ to be divisible by 504 is $789,264 + 504 = 789,768$, which repeats both the 7 and the 8.

60. ONE PIECE FEWER

Cutting the 8×8 checkerboard into three pieces will never be enough. The easiest way to see that four pieces are required is to note that the four corner squares of the 8×8 checkerboard must each belong to a different piece in the dissection. Why? Because a piece containing two corner squares could never fit into a square smaller than 8×8!

61. SCALE DRAWING

The equation linking the two scales is $F = (9/5)C + 32$. To find the temperature that reads the same on both scales, just set F and C equal to one another, yielding $C = (9/5)C + 32$. This gives $(4/5)C = -32$, so $C = -40$. Therefore the only temperature that reads the same on both the Celsius and Fahrenheit scales is 40 degrees below zero.

62. PLAYING THE TRIANGLE

The dissection can be completed as follows: The length of a side of triangle A must be ⅗ of a side of the original triangle, and a side of the three-piece triangle is ⅘ the original. The key is to cut the triangle in such a way that the shorter sides of piece B (bottom and right) are equal.

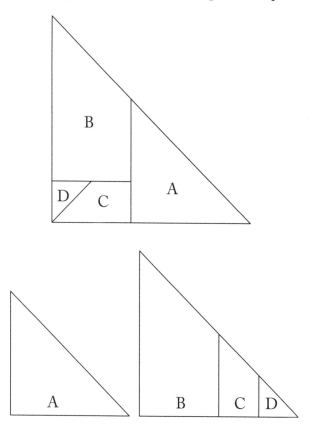

63. HOW BIG?

The easiest method is to draw a triangle within the wedge, as shown on the next page.

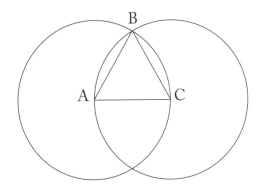

Note that triangle ABC must be equilateral, because each side is a radius of the circle, so angle BAC is 60 degrees. The pie-shaped figure determined by angle BAC (or angle BCA) must therefore have area $\pi/6$, because it represents one-sixth of a circle of radius one. The area of the upper half of the wedge we're interested in is the sum of those two sectors minus the area of the triangle, or $2(\pi/6) - \sqrt{3}/4$. The whole wedge thus has area $2\pi/3 - \sqrt{3}/2$.

64. TUNNEL DIVISION

There are 5,280 feet in a mile and 3,600 seconds in an hour (3,600 = 60 × 60). Therefore, 90 miles per hour is the same as $(90 \times 5280)/3600 = (9 \times 528)/36 = 528/4 = 132$ feet per second.

With this in mind, if the train takes four seconds to completely enter the tunnel, its length must be 132 × 4 = 528 feet. The length of the tunnel must be 132 × 40, or 5,280 feet. The tunnel is precisely one mile long!

65. PRIME TIME

There are many different solutions. One, for example, is given by the sequence 0 7 4 3 2 5 6 1.

66. TRIANGLE EQUALITIES

The three triangles are as follows:

Triangle	Area	Perimeter
12–16–20	96	48
10–24–26	120	60
9–40–41	180	90

Note that the first triangle is four times the 3–4–5 right triangle, while the second one is twice the 5–12–13 right triangle.

67. DIVIDING THE PENTAGON

The solution is reached by forming another regular pentagon within the larger one, rotating the smaller one slightly, then joining the endpoints. The result is this figure:

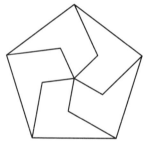

68. PLANTING THE SOD

Although 4444^{4444} is an enormous number, it gets substantially smaller upon taking the sum of its digits. To begin with, the number of digits in 4444^{4444} is less than the number of digits in 10000^{4444} (which equals $(10^4)^{4444} = 10^{17776}$), which has 17,777 digits. So $\text{SOD}(4444^{4444}) < 17,777 \times 9 = 159,993$. Therefore $\text{SOD}(\text{SOD}(4444^{4444})) < \text{SOD}(99,999)$, because the sum of the digits in 99,999 is at least as big as the SOD of any other number less than 159,993. Now $\text{SOD}(99,999) = 45$, so $\text{SOD}(\text{SOD}(\text{SOD}(4444^{4444})))$ is at most SOD(39)—again, 39 has the highest SOD of any number less than 45—and this equals 12.

But we also know that if we divide SOD(SOD(SOD $(4444^{4444})))$ by 9, the remainder is the same as the remainder upon dividing 4444^{4444} by 9, and, believe it or not, this can be calculated. We know that $4444 \equiv 7$ (mod 9), and that $7^3 = 343 \equiv 1$ (mod 9), so $4444^{4444} \equiv 7^{4444} = 7 \times 7^{4443} = 7 \times (7^3)^{1481} \equiv 7$ (mod 9). Therefore SOD(SOD(SOD(4444^{4444}))) is a number that is less than 12 and is congruent to 7 (mod 9), so it must be 7.

69. HIGHER THAN YOU THINK

The smallest number N such that it is impossible to create a dollar out of N coins is N = 77. That's because it is impossible to create 25 cents out of two coins; more generally, it is impossible to create 25 + 5x cents out of 2 + 5x coins.

70. POCKET CHANGE

Your friend must have 15 coins in his pocket. Here are the six ways that you can create a dollar from 15 coins. (Remember, no half-dollars allowed.)

	Q	D	N	P
1)	3	1	1	10
2)	2	1	7	5
3)	1	1	13	0
4)	1	5	4	5
5)	0	9	1	5
6)	0	5	10	0

Believe it or not, 15 is the only number such that there are precisely six ways of creating a dollar from that number of coins! The answer to the puzzle is therefore unique.

71. SQUARE NOT

128 is the largest integer that cannot be expressed as the sum of distinct squares. Note that 128 − 100 = 28, which is not expressible, and similarly for 128 − 81 = 47, etc. For the record, here are some ways in which the integers from 129 through 150 can be expressed:

129 = 100 + 25 + 4
130 = 100 + 25 + 4 + 1
131 = 81 + 49 + 1
132 = 81 + 25 + 16 + 9 + 1
133 = 81 + 36 + 16
134 = 81 + 36 + 16 + 1
135 = 100 + 25 + 9 + 1
136 = 100 + 36
137 = 100 + 36 + 1
138 = 100 + 25 + 9 + 4
139 = 100 + 25 + 9 + 4 + 1

140 = 100 + 36 + 4
141 = 100 + 36 + 4 + 1
142 = 100 + 25 + 16 + 1
143 = 81 + 36 + 25 + 1
144 = 144
145 = 144 + 1
146 = 100 + 36 + 9 + 1
147 = 81 + 36 + 25 + 4 + 1
148 = 144 + 4
149 = 100 + 49
150 = 100 + 49 + 1

72. DUELING WEATHERMEN

If it is sunny, that means that WET was wrong and WILD was right, which occurs with probability = $(\frac{1}{4})(\frac{4}{5}) = \frac{1}{5}$.

If it rains, WET was right and WILD was wrong, which occurs with probability = $(\frac{3}{4})(\frac{1}{5}) = \frac{3}{20}$.

The odds of rain are therefore $\frac{3}{20}$ to $\frac{1}{5}$, or 3 to 4. Therefore the probability of rain equals $\frac{3}{(3 + 4)} = \frac{3}{7}$.

73. TWO-WAY ADDITION

The addition at right works both right side up and upside down.

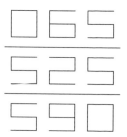

74. THE ICING ON THE CAKE

Although the diagram below isn't drawn completely to scale, it shows one way of cutting the cake so as to give each kid the same amount of cake and frosting. Note that the middle wedge contains an 8 × 12 section on top—96 square inches of frosting—plus two triangles on the sides, each with $\frac{(8 \times 6)}{2} = 24$ square inches of frosting, for a total of 144 square inches, as desired. The other two pieces must have the same amount of frosting, so they must split the remaining 288 square inches of frosting right down the middle.

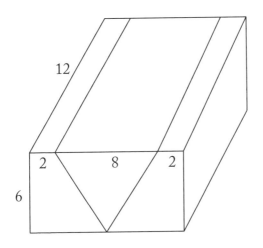

As for the amount of cake, if you cut straight down, the piece that is eight inches long would be four times as big as each of the other two pieces. By cutting at an angle, you transfer one-fourth of the volume of the big piece to each of the smaller pieces, so they end up being exactly the same size. Other answers are possible.

75. HIGH MATH AT THE 7-11

Note that 711 = 9 × 79, and 79 is prime. One of the items must therefore be a multiple of $0.79. Here are the first six multiples of $0.79, followed by $7.11 minus these numbers (the sum of the remaining three items) and $7.11 divided by these numbers (the product of the remaining three items).

	0.79	1.58	2.37	3.16	3.95	4.74
Sum	6.32	5.53	4.74	3.95	3.16	2.37
Prod.	9.00	4.50	3.00	2.25	1.80	1.50

We now must solve the problem of what *three* numbers add to the sum and multiply to the corresponding product. Only the pair (3.95, 2.25) looks at all friendly, because the others force us to create a product that is a round number from a sum that is not round. After a little trial and error, note that if you subtract (divide) the sum (product) number by 1.25, you get a required sum (for two numbers) of 2.70, and a corresponding product of 1.80. Now we're talking. We see that 1.50 and 1.20 add to 2.70 and multiply to 1.80. The four items therefore cost $3.16, $1.25, $1.50, and $1.20.

76. WHO AM I?

From statement 2, if I'm a multiple of 3, I must be one of 51, 54, or 57. But none of these is a multiple of 4, which contradicts statement 1. So I'm not a multiple of 3. If I'm not a multiple of 3, I can't be a multiple of 6, so by statement 3, I must be one of 71, 73, 74, 75, 76, 77, 79. But if I'm one of these numbers, then statement 1 says that I must be a multiple of 4. And the only multiple of 4 in that list is 76. So I am 76.

77. HOME ON THE RANGE

If 11 sheep can last 8 days, it would appear that the pasture has 88 "sheep-days" in it. But if 10 sheep can last 9 days, for a total of 90 sheep-days, we have to conclude that the extra day's growth amounts to two sheep-days. If so, two sheep could last their entire lives!

78. PARTY OF 12

To maintain a pattern of alternating genders, something has to give. Either you put six people on one side of the table and four on the other (not that desirable), or you place two people at either end, as in the diagram below.

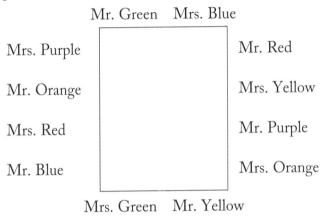

79. HEADS OR TAILS

The probability is precisely ½ that at some point there will be 3 or more consecutive flips that come out the same.

To see why, look first at the probability that three or more heads will come up. There are eight sequences with three consecutive heads, as follows: HHHTT, HHHTH, HHHHT, THHHT, THHHH, HTHHH, TTHHH, HHHHH.

Similarly, there are eight sequences with three or more tails. Putting them together, 16 of the 32 possible combinations involve three consecutive flips that are the same, so the probability of this event equals ½.

80. AS EASY AS PI

Let r equal the radius of the circle, which is the same as one-half the side of the large square. Then the area of each smaller square equals r^2, so the total area of the three shaded squares equals $3r^2$. But the area of the circle equals πr^2. Because π is greater than 3, the circle must represent the larger shaded region.

81. TAKING THE LONG SHOT

First we must translate the odds into probabilities. Note that the chance of Lightning Bolt's winning is ⅓; similarly, Golden Honey has a ¼ chance of winning, and Matchmaker a ⅕ chance.

We seek amounts A, B, and C such that $A + B + C + 39 = 3A = 4B = 5C$. Therefore, $B = (¾)A$ and $C = (⅗)A$. Substituting gives $(¾)A + (⅗)A + 39 = 2A$, so $39 = (¹³/₂₀)A$, so $A = 60$. Similarly, $B = 45$, and $C = 36$, so we must bet $60 on Lightning Bolt, $45 on Golden Honey, and $36 on Matchmaker.

82. TRICK TRIG QUESTION

The equation is always true, provided that you allow for some creative cancellation:

$$\frac{\sin(x)}{n} = \frac{si\cancel{n}(x)}{\cancel{n}} = si(x) = six$$

83. MAPPING IT OUT

The maximum number of lines that can emanate from any one city is five. To see why, suppose one city—call it city A—has six lines emanating from it. Then there must exist two cities B and C such that angle BAC is less than 60

82

degrees. (If all angles were precisely 60 degrees, the proof would be similar.) Now we use the fact that the shortest side of a triangle is the one opposite the smallest angle. If angle BAC is smaller than 60 degrees, then at least one of angles ABC and ACB is greater than 60 degrees. But then segment BC is shorter than AB or AC (or both), contradicting the fact that A is the closest city to B and C. Therefore five cities is the maximum.

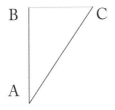

84. CLOSE COUNTS IN HORSESHOES

The following arrangement works:

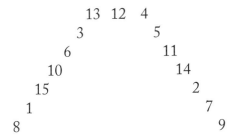

Note that the 8 and the 9 must be on the bottom of the horseshoe, because each of these numbers only has one other number within the 1–15 range that will sum to a square. (8 + 8 = 16, a square, but numbers cannot be repeated.)

85. SEE YOU LATER, ESCALATOR

The escalator always has 50 steps showing. To see why, suppose that the escalator moves at 1 step per second. (This is an arbitrary selection, but convenient.) Let n represent the number of steps showing at any one time on the escalator. When Frank descends the escalator, he walks 25 steps, which means the movement of the escalator carries him the other $(n - 25)$ steps. Since we chose that the escalator moves at a rate of one step per second, Frank is on the escalator for a total of $(n - 25)$ seconds. So Frank's stepping rate is $\frac{25}{(n-25)}$ steps per second. Using similar logic, Dave's stepping rate is $\frac{20}{(n-20)}$ steps per second. We're told that for every two steps Dave takes, Frank takes three. So $2(\frac{25}{(n-25)}) = 3(\frac{20}{(n-20)})$. This simplifies to $\frac{5}{(n-25)} = \frac{6}{(n-20)}$. Cross-multiply and rearrange to get that $n = 50$. Since the escalator's rate does not affect the solution, we see that the escalator always has 50 steps showing.

86. WHITE ON RED

If there are W white socks and S socks altogether, then the number of ways of choosing two socks is $\frac{S(S-1)}{2}$. The number of ways of choosing two white socks equals $\frac{W(W-1)}{2}$. The probability of choosing two white socks in succession therefore equals $\frac{W(W-1)}{S(S-1)}$.

To solve the problem, we must therefore find two pairs of consecutive integers, with the product of the first pair being precisely one-third the product of the second pair. By using a little trial and error, we see that we are in luck:

$$2 \times 3 = 6$$
$$3 \times 4 = 12$$
$$4 \times 5 = 20$$
$$5 \times 6 = 30$$
$$6 \times 7 = 42$$
$$7 \times 8 = 56$$

$8 \times 9 = 72$

$9 \times 10 = 90$

Since $9 \times 10 = 90 = 3 \times 30 = 3 \times 5 \times 6$, there must be 10 socks altogether—six white and four red.

87. THE 18-12 OVERTURE

Let A be the number in the lower right corner, and B be the number in the lower left corner. Then the number between 12 and B in the first column must be A + 6. We can fill in the remainder of the square as follows:

12	A−6	B+12
A+6	B+6	6
B	18	A

Summing the upward sloping diagonal and equating it to the remaining row, column, and diagonal sums, we get the equation 3B + 18 = A + B + 18, so A = 2B. Setting A = 20, for example, yields the solution at right.

12	14	22
26	16	6
10	18	20

88. SITTING ON THE FENCE

A flat rate of $15 is the better option, not only because it is more reliable, but because it has the higher expected payoff.

The expected payoff of the second option equals the sum of the possible dollar receipts multiplied by their probabilities. This calculation gives us $2(1/15) + $6(4/15) + $10(1/15) + $11(2/15) + $15(2/15) + $21(2/15) + $25(2/15) + $30(1/15) = $210/15 = $14. The flat rate therefore has an expected payoff of $1 more per week.

89. TURNING ON A DIME

The correct answer is five. The distance the coin rolls is equal to the distance traveled by its center. The center travels in a path that forms a circle of radius 5. The center travels 10π units, so the smaller coin rolls through 10π units. Since this smaller coin has a circumference of 2π units, it makes 5 full revolutions. In general, if the circumference of the larger circle is N times that of the smaller circle, the number of complete revolutions equals N + 1. (Note that if the two circles are the same size, the number of revolutions is two, not one.)

90. RAZOR-SHARP REASONING

The total distance around equals $3 + \pi$ inches. That's because each of the three arcs consists of one-third the circumference of a circle, and each of the three straight segments consists of two radii linked together, so each must be one inch long.

91. FAIR AND SQUARE

A = 2, B = 4, C = 5, D = 6, and E = 9. Note that 4 + 16 + 25 + 36 = 81.

92. SPECIAL DELIVERIES

Sunnyside Hospital has the greater chance. In general, the greater the number of deliveries, the less the chance that they will divide precisely evenly (assuming an even number, of course). Note that if a hospital delivered precisely two babies, the probability of one boy and one girl would be one-half. For 200 babies, you could only obtain a probability of ½ if the first 199 were divided 100 to 99, meaning that the likelihood of an even division is considerably less than one-half. (In general, the probability of an even division for 2N births equals $(2N)!/N!N!2^{2N}$, and this number goes down as N goes up.)

93. BATTLE STATIONS

Let x represent the distance the battalion moves forward in the time the rider goes from back to front. While the rider travels (20 + x) miles, the battalion travels x. Similarly, as the rider rides back, the rider travels x and the battalion travels (20 − x). But the ratio of rider speed to battalion speed remains constant throughout, so $(20 + x)/x = x/(20 - x)$. Clearing fractions, we have $x^2 = 400 - x^2$, so $2x^2 = 400$, so $x = 10\sqrt{2}$. The messenger traveled a total distance of 20 + 2x, which is $20 + 20\sqrt{2}$.

94. SPENDING SPREE

Some educated guesses get this one done, beginning with the inequalities 2P < B, 3P > 4C, and 3C > B. It turns out that the solution is P = $26, C = $19, and B = $55.

95. A TWO-AND-ONE COUNT

The formula for the sum of the first N integers equals $\frac{N(N+1)}{2}$. Therefore the first step is to find the biggest integer such that $N(N+1) < 4{,}000$. This number is 62, because $62 \times 63 = 3{,}906$, so the sum from 1 to 62 must equal $\frac{3906}{2} = 1{,}953$. This is 47 less than 2,000, so the number 47 must have been counted twice.

96. RATIONAL THINKING

The equation of the upward sloping line is simply $y = x$, while the equation of the downward sloping line is given by $y = -\frac{2}{3}(x - 10)$, or simply $2x + 3y = 20$. (Note that both $(4,4)$ and $(10,0)$ satisfy this equation.) The upper-left corner of the square is on the line $y = x$, so its coordinates are (n,n), where n is the length of a side of the square. The coordinates of the upper-right corner are then $(2n,n)$. By plugging this point into the second equation, we get $2(2n) + 3n = 20$, or $7n = 20$. Therefore the length of a side of the square equals $\frac{20}{7}$.

97. ROAMIN' CANDLES

Suppose for convenience that each candle is $4 \times 7 = 28$ inches tall—even though candles aren't often that tall in real life. After one hour, candle A will be 21 inches and candle B will be 24 inches—not enough. After two hours, A will be 14 inches and B will be 20 inches—still not enough. After three hours, A will be 7 inches and B will be 16 inches—too much! So clearly the moment at which B is twice A will be between two and three hours from the start.

Let's get the precise answer. The fraction of candle A that remains after t hours is $\frac{(4-t)}{4}$ and the fraction of candle B that remains is $\frac{(7-t)}{7}$. We want to find the point at which the slower-burning candle is twice as tall as the faster-burning candle. This happens when $2(^{(4 - t)}\!/_4) = {}^{(7 - t)}\!/_7$, so $28 - 7t = 14 - 2t$, which gives us that $5t = 14$, or $t = 2.8$ hours. This is the same as 2 hours and 48 minutes.

Note that the choice of 28 inches for the candle heights has no bearing on the algebraic solution. The answer $T = 2.8$ hours will work for any starting height.

98. I'M EIGHTEEN, AND I LIKE IT

If a number n is factored as $n = P_1^{Q_1} \times P_2^{Q_2} \times P_3^{Q_3} \dots P_K^{Q_K}$, then the number of factors of n (including 1 and n) equals $(Q_1 + 1)(Q_2 + 1)(Q_3 + 1) \dots (Q_N + 1)$.

In this case, the total number of factors of n must equal 20 (18 proper factors plus 1 and n), and the smallest number with this property is $2^4 \times 3 \times 5 = 240$.

For the record, the 18 proper factors of 240 are 2, 3, 4, 5, 6, 8, 10, 12, 15, 16, 20, 24, 30, 40, 48, 60, 80, and 120.

99. MILES TO GO

Jones must drive an additional 861.1 miles, at which point the main odometer will read 13206.7 miles and the trip odometer will read 984.5 miles. (I believe this puzzle dates back to a construction of Harry Nelson.)

100. UPSIDE-DOWN PYTHAGORAS

The smallest solution is $x = 15$, $y = 20$, and $z = 12$.

101. THE ABSENT-MINDED PROFESSOR

This problem is best handled geometrically. The probability of the two twins' scores coming within 20 of one another is simply the area of the shaded region below divided by the area of the entire square. The two non-shaded pieces, if put together, form a square whose side is ⅘ the length of the original square. Therefore the ratio in question is $1 - (⅘)(⅘) = 1 - 16/25 = 9/25$.

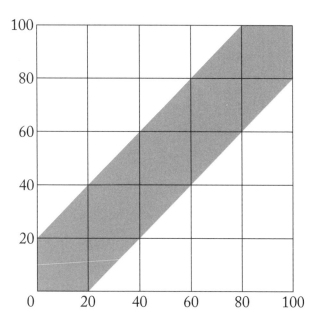

Index

Answer page numbers are in italics.

About the Author

DERRICK NIEDERMAN is the author of *Number Freak, The Puzzler's Dilemma*, and several volumes of math puzzles and brainteasers. He has also invented a variety of games and puzzles, including 36 Cube and PathWords. He lives in Charleston, South Carolina, and teaches mathematics at the College of Charleston.